the [illegible]e
g[illegible]e
revolution
for people with diabetes

DR STEPHEN COLAGIURI
KAYE FOSTER-POWELL • DR JENNIE BRAND MILLER

CORONET BOOKS
Hodder & Stoughton

First published in Australia in 1997 by
Hodder Headline Australia Pty Limited

First published in Great Britain in 2000
by Hodder and Stoughton
A division of Hodder Headline

This United Kingdom edition is published by arrangement with Hodder
Headline Australia Pty Limited
A Coronet Paperback

10 9 8 7 6 5

A CIP catalogue record for this title is available from the
British Library

ISBN 0 340 76989 0

Printed and bound in Great Britain by
Clays Ltd, St Ives plc

Hodder and Stoughton
A division of Hodder Headline
338 Euston Road
London NW1 3BH

A word of advice
There are many factors that can affect your blood sugar levels. If you have diabetes and you are struggling to control your blood sugar level it is important to seek medical help. How much exercise you do, your weight, stress levels, total dietary intake and need for medication may have to be assessed.

CONTENTS

INTRODUCTION

Health authorities all over the world stress the importance of high carbohydrate diets for good health and diabetes management. The question now is which type of carbohydrate is best for people with diabetes? Research on the glycaemic index (what we call the G.I. factor) shows that different carbohydrate foods have dramatically different effects on blood sugar levels.

The G.I. factor gives you the true story about carbohydrate and the blood sugar connection. For people with diabetes this can mean a new lease of life. Literally!

Understanding the G.I. factor has made an enormous difference to the diet and lifestyle of people with diabetes because the findings reveal that:

- many traditionally 'taboo' foods don't cause the unfavourable effects on blood sugar they were believed to have;
- diets with a low G.I. improve blood sugar control in people with insulin-dependent and non-insulin dependent diabetes;
- many more foods make up a healthy diet for someone with diabetes than was once believed.

WHY WE WROTE THIS SPECIAL POCKET BOOK

In *The Glucose Revolution* we broadcast the fact that there are different types of carbohydrate that work in different ways. Here, we show you how the G.I. factor can help you achieve better diabetes control because:

- it is a proven guide to the true physiological effects of foods on blood sugar levels; and
- it provides an easy and effective way to eat a healthy diet and control fluctuations in blood sugar.

The G.I. factor is a ranking of foods based on their immediate effect on blood sugar levels.

- Carbohydrate foods that break down quickly during digestion have the highest G.I. factors. Their blood sugar response is fast and high.
- Carbohydrate foods that break down slowly, releasing glucose gradually into the bloodstream, have low G.I. factors.

The rate of carbohydrate digestion has important implications for everybody. It's vital that people with diabetes learn about the G.I. factor so they base their diet on sound scientific evidence.

HOW TO USE THIS BOOK

This book is a basic guide to the G.I. factor and diabetes and gives practical examples to help you put it into practice.

Many people with diabetes find that despite doing all they are told, their blood sugar levels remain too high. This book contains the most up-to-date information about carbohydrate and the optimum diet for people with diabetes. It explains which types of carbohydrate are best and why – information based on scientific research, clinical trials and the real experiences of real people.

- We show you how to include more of the right sort of carbohydrate in your diet.
- We provide practical hints for meal preparation and tips to help you make the G.I. factor work for you throughout the day.
- We give a week of low G.I. menus plus a nutritional analysis for each menu and its G.I. factor.
- We explain how the G.I. factor is measured.
- We include an A to Z listing of over 300 foods with their G.I. factor, carbohydrate and fat count.

WHAT DOES THE G.I. FACTOR MEAN FOR PEOPLE WITH DIABETES?

One of the major aims of diabetes therapy is to maintain near normal blood sugar levels. Not long ago, people with diabetes were told to eat complex carbohydrates (starches) because it was believed they were slowly absorbed causing a smaller rise in blood sugar levels. Simple sugars were restricted because they were thought to be quickly absorbed and their blood sugar response would therefore be fast and high.

These assumptions were wrong. We now know that the concept of simple and complex carbohydrates does not tell us how carbohydrate will actually behave in the body. Different carbohydrate-containing foods do have different effects on blood sugar levels, but we can't predict the effect by looking at the sugar or starch content.

Since the 1980s scientists have studied the actual blood sugar responses to hundreds of different foods on healthy people and people with diabetes. They gave them real foods and then measured the blood sugar levels at frequent intervals, for up to two to three hours after the meal. To compare foods according to their true physiological effect on blood sugar levels, they came up with the term 'glycaemic index' (what we call the G.I. factor). This is simply a ranking of foods from 0 to 100 that tells us whether a food will raise blood sugar levels dramatically, moderately, or just a little.

Research on the G.I. factor has turned some widely held beliefs upside down.

- The first surprise was that many starchy foods (bread, potatoes and many types of rice) are digested and absorbed very quickly, not slowly as had always been assumed.
- The next surprise was that moderate amounts of many sugary foods did not produce dramatic rises in blood sugar as had always been thought.

The truth was that many foods containing sugar actually showed intermediate blood sugar responses, often lower than foods like bread. So, forget the old distinctions that were made between starchy foods and sugary foods or simple versus complex carbohydrate. These distinctions have little scientific basis. The G.I. factor tells another story closer to the truth.

It's time to forget about
simple and complex carbohydrate
and to think in terms of
low G.I. and high G.I. factor.

WHY IS THE G.I. FACTOR SO IMPORTANT IN DIABETES?

If blood sugar levels are not properly controlled, diabetes can cause damage to the blood vessels in the heart, legs, brain, eyes and kidneys. For this reason, heart attacks, strokes, kidney failure and blindness are more common in people with diabetes. High blood sugar levels can also damage the nerves in the feet causing pain and irritation in the feet and numbness and loss of sensation.

Damage to the blood vessels of the heart, legs and brain may also be caused by high levels of insulin. Some researchers think that high insulin levels might stimulate the muscle in the wall of the blood vessel to thicken. This thickening causes the blood vessels to narrow and can slow the flow of blood to the point that a clot forms and stops the blood flow altogether. This is what causes a heart attack or stroke.

In general, studies show an excellent correlation between the G.I. factor of a food and its insulin response. With low G.I. foods, there's a reduced secretion of the hormone insulin over the course of the day. With high G.I. foods, the body produces larger amounts of insulin, resulting in higher levels of insulin in the blood.

It makes sense for people with Type 2 or non-insulin dependent diabetes to eat foods with a low G.I. factor to help control blood sugar levels, and do so with lower levels of insulin. This may have the added benefit of reducing the large vessel damage which accounts for many of the problems that diabetes can cause.

A low G.I. diet improves the body's sensitivity to insulin, so the insulin you do have works better.

We also know that a low G.I. diet in conjunction with a low fat intake can help keep your blood vessels healthy by keeping your blood fat levels down. Studies have shown that raised blood fat levels (like cholesterol and triglycerides) are reduced in people who reduce the G.I. factor of their diet.

The slow digestion and gradual rise and fall in blood sugar after a low G.I. food help control blood sugar levels in people with diabetes.

WHAT IS THE OPTIMUM DIET FOR PEOPLE WITH DIABETES?

For over a hundred years, people with diabetes have been given advice on what to eat. Many diets were based more on unproven (although seemingly logical) theories, rather than actual research. In 1915, for example, the *Boston Medical and Surgical Journal* advocated that the best dietary treatment for someone with diabetes was 'limitation of all components of the diet'. This translated to a very low kilojoule diet interspersed with days of fasting. Unfortunately, malnutrition was often the result!

Fortunately, good quality scientific research supports today's dietary recommendations for people with diabetes. We now know that:

A diet which is good for people with diabetes is a diet that is good for everyone.

The optimum diet for people with diabetes contains a wide variety of foods.

CHECKLIST: THE OPTIMUM DIET FOR PEOPLE WITH DIABETES

Plenty of wholegrain cereals, breads, vegetables and fruits

A low fat, low G.I. diet contains lots of heavy grain breads; cereals like rice, barley, couscous, cracked wheat; legumes like kidney beans and lentils; and all types of fruit and vegetables.

Only small amounts of fat, especially saturated fat

Limit biscuits, cakes, butter, potato chips, takeaway fried foods, full cream dairy products, fatty meats and sausages which are all high in saturated fat. Poly- and mono-unsaturated oils are healthier types of fats.

A moderate amount of sugar and sugar containing foods

It's OK to include your favourite sweetener or sweet food – small quantities of sugar, honey, golden syrup, jam – to make meals more palatable and pleasurable.

Only a moderate quantity of alcohol

Only 3 standard drinks for men and 2 standard drinks for women per day.

Only a moderate amount of salt and salted foods.

Try lemon juice, fresh ground black pepper, garlic, chilli, herbs and other flavours rather than relying on salt.

HOW DOES FOOD AFFECT OUR BLOOD SUGAR?

Our bodies burn fuel all the time and the fuel our bodies like best is carbohydrate. Carbohydrate is the *only* fuel that the brain and red blood cells can use.

When we eat carbohydrate foods, the body breaks them down in the gut into a form that can be absorbed and which the cells can use. This process is called digestion. Digestion starts in the mouth when amylase, the digestive enzyme in saliva, is incorporated into the food by chewing. The activity of this enzyme stops in the stomach. Most of the digestion continues only when the carbohydrate reaches the small intestine.

In the small intestine, amylase from pancreatic juice breaks down the large molecules of starch into short chain molecules. These and any disaccharide sugars are then broken into simpler monosaccharides by enzymes in the wall of the intestine. The monosaccharides that result, glucose, fructose and galactose, are absorbed from the small intestine into the bloodstream.

Carbohydrate is the only part of food that directly affects blood sugar levels.

The body responds by releasing insulin into the blood which clears the sugar from the blood, moving it into the muscles and cells where it is used for energy. Some glucose is maintained in the blood to serve the brain and central nervous system.

There are still people who think that because carbohydrate raises blood sugar, people who have diabetes should not eat it at all. This is wrong. Carbohydrate is a necessary part of a healthy diet. For people with diabetes, choosing carbohydrate foods with a low G.I. lessens the rise and fall in blood sugar and helps achieve more stable blood sugar levels.

- People with insulin-dependent diabetes need to balance the amount and timing of carbohydrate in their diet with their dose of insulin and their activity level (rarely an easy task).
- People with non-insulin dependent diabetes, who have a relative lack of insulin, should distribute their carbohydrate intake across the day and may need to consider the timing of their meals in relation to diabetic tablets they take.

At least half of our total daily kilojoules should come from carbohydrate.

DISPELLING SOME MYTHS ABOUT FOOD AND DIABETES

Research into the glycaemic index shows that some of the popular beliefs about food and diabetes are not true.

Myth 1: Sugar is the worst thing for people with diabetes.

Not true. Sugar and sugary foods in normal serves have no greater effect on blood sugar levels than many starchy foods. Fat is far worse for people with diabetes.

Myth 2: All complex carbohydrates or starches are slowly digested in the intestine.

Not true. Some starch, like that in potatoes, is digested quickly, causing a greater change in blood sugar level than many sugar-containing foods.

Myth 3: Starchy foods like bread and potatoes are fattening.

Not true. Bread and potatoes are carbohydrate (fuel) foods which are the foods your body burns most readily. They are the least likely to be stored as body fat.

Myth 4: Eating a lot of sugar causes diabetes.

Not true. A diet high in fat and quickly digested carbohydrates contributes to obesity which makes non-insulin dependent diabetes more likely to appear in those who are at risk.

Myth 5: You can't lose weight eating between meal snacks.

Not true. The type and total amount of kilojoules consumed and the amount of kilojoules the body uses determine body weight. Low fat, high carbohydrate snacks can safely be included in a low kilojoule eating plan.

Myth 6: Sugar is fattening.

Not true. Sugar is just another carbohydrate, and it's almost impossible to turn it into body fat.

Myth 7: High blood sugar is caused by eating too much sugar.

Not true. A number of factors can cause blood sugar levels to rise including the body's response to stress or illness, reduced activity, missed medications and excess carbohydrate.

Myth 8: Sugar in the diet will result in lower intakes of vitamins and minerals.

Not true. Studies show that diets containing moderate amounts of refined sugars are perfectly healthy (10 to 12 per cent of kilojoule intake) and the sugar helps make many nutritious foods like porridge more palatable. Diets high in natural sugar from a range of sources including dairy foods and fruit often have higher levels of micro-nutrients such as calcium, riboflavin and vitamin C.

5 KEYS TO A HEALTHY DIET

- Eat carbohydrate-rich foods at every meal and make sure that carbohydrates and vegetables form a large proportion of the meal.
- Eat carbohydrate-rich foods for snacks, rather than high fat foods.
- Include at least the minimum quantity of carbohydrate foods suggested for small eaters (see page 20).
- Make at least half your carbohydrate choices, foods with a low G.I. factor.
- Do not eat too much fat. High fat foods are a concentrated source of Calories. It takes only a little extra of them to throw your diet out of balance.

A healthy diet is high in carbohydrate and low in fat.

To improve the quality of our diet most of us need to eat more carbohydrate and less fat.

HOW MUCH CARBOHYDRATE DO YOU NEED?

About half of our total Calorie intake should come from carbohydrate. We need to consume 150 g of carbohydrate for every 1000 Calories.

For a low Calorie diet (1200 Calories), it means eating about 190 g of carbohydrate per day (equivalent to 14 slices of bread).

For a young, active person with higher energy requirements, say in the order of 2000 Calories, it means eating 300 g of carbohydrate per day (equivalent to 24 slices of bread).

On page 20 for small eaters, and page 21 for bigger eaters, we have calculated sample carbohydrate intakes.

Your Calorie (and carbohydrate) requirements depend on your age, gender, activity level and body size. It is not possible to publish standard figures that will apply to everybody. If you want to assess your own specific Calorie requirements and calculate exactly how much carbohydrate you need, we suggest that you consult a dietitian.

Once you have the amount of carbohydrate in your diet right, the next step is to choose the right type of carbohydrate foods, i.e., those with a low G.I. factor.

CARBOHYDRATE REQUIREMENTS FOR SMALL EATERS

You might consider yourself a small eater if you:

- are a small-framed female
- have a small appetite
- do very little physical activity
- are trying to lose weight

Even the smallest eater needs these carbohydrate foods every day. This food list supplies 225 g of carbohydrate, suitable for a 1500 Calorie diet.

- around 4 slices of bread or the equivalent (crackers, rolls)
  PLUS
- about 3 pieces of fruit or the equivalent (fresh, canned, dried)
  PLUS
- 1 cup of high carbohydrate vegetables (corn, legumes, potato, sweet potato)
  PLUS
- at least 1 cup of cereal or grain food (breakfast cereal, cooked rice or pasta, or other grains)
  PLUS
- 2 cups of low fat milk or the equivalent (yoghurt, ice cream). This includes milk in your tea and coffee and with your cereal

CARBOHYDRATE REQUIREMENTS FOR BIGGER EATERS

The picture of a bigger eater would fit you if you are:

- an active young female of average frame size
- doing regular physical activity (but not prolonged strenuous exercise)
- an active adult male or teenage boy
- working as a labourer

The following food list provides 375 g of carbohydrate which is suitable for a 2500 Calorie diet.

- around 8 slices of bread or the equivalent (crackers, rolls)
  PLUS
- about 3 pieces of fruit or the equivalent (fresh. canned, dried)
  PLUS
- 2 cups of high carbohydrate vegetables (corn. legumes, potato, sweet potato)
  PLUS
- at least 2 cups of cereal or grain food (breakfast cereal or cooked rice, or pasta or other grain)
  PLUS
- 2 cups of low fat milk or the equivalent (yoghurt, ice cream).This includes milk in your tea and coffee and with your cereal

IS IT BETTER TO EAT COMPLEX CARBOHYDRATE INSTEAD OF SIMPLE SUGARS?

There are no big distinctions between sugars and starches in either nutritional terms or in the G.I. sense. Some sugars such as fructose or fruit sugar have a low G.I. factor. Others, such as glucose, have a high G.I. factor. The most common sugar in our diet, ordinary table sugar (sucrose), has a moderate G.I. factor.

Starches can fall into both the high and low G.I. categories too, depending on the type of starch and what treatment it has received during cooking and processing. Most modern starchy foods, like bread, potatoes and breakfast cereals, contain high G.I. carbohydrate.

What our research has shown is that you don't have to eliminate sugar completely from your diet. However, it is important to remember that sugar alone won't keep the engine running smoothly, so don't overdo it.

A balanced diet contains a wide variety of foods.

ARE NATURALLY OCCURRING SUGARS BETTER THAN REFINED SUGARS?

Naturally occurring sugars are those found in foods like fruit, vegetables and milk. Refined sugars are concentrated sources of sugar such as table sugar, honey or molasses.

The rate of digestion and absorption of naturally occurring sugars is no different, on average, from that of refined sugars. There is wide variation within both food groups, depending on the food. For example, the G.I. of fruits ranges from 22 for cherries to 72 for watermelon. Similarly, among the foods containing refined sugars, some have a low G.I. and some have a high G.I. factor. The G.I. of sweetened yoghurt is only 33, while jelly beans have a G.I. of 80.

Some nutritionists argue that naturally occurring sugars are better because they contain minerals and vitamins not found in refined sugar. However, recent studies which have analysed high sugar and low sugar diets clearly show that the diets overall contain similar amounts of micronutrients. Studies have shown that people who eat moderate amounts of refined sugars have perfectly adequate micronutrient intakes.

DOES SUGAR CAUSE DIABETES?

No. There is absolute consensus that sugar does not cause diabetes. Contrary to popular belief, it is an excess of fat in the diet, leading to an excess of body fat, which is more heavily implicated in the development of some diabetes.

Type 1 diabetes (insulin-dependent diabetes) is an autoimmune health problem triggered by unknown environmental factors possibly such as viruses.

Type 2 diabetes (non-insulin dependent diabetes) is strongly inherited but lifestyle factors such as lack of exercise and overweight increase the risk of developing it. Because the dietary treatment of diabetes in the past involved strict avoidance of sugar, many people wrongly believed that sugar was in some way implicated as a cause of the disease.

There is absolute consensus that sugar does not cause diabetes.

CAN PEOPLE WITH DIABETES EAT AS MUCH SUGAR AS THEY WANT?

Research shows that moderate consumption of refined sugar (around 40 g or 2 tablespoons) a day doesn't compromise blood sugar control. This means you can choose foods which contain refined sugar or even use small amounts of table sugar. Try to spread your sugar budget over a variety of nutrient rich foods that sugar makes more palatable. Remember, sugar is concealed in many foods – a can of soft drink contains about 40 g sugar.

Most foods containing sugar do not raise blood sugar levels any more than most starchy foods. Kelloggs Cocopops™ (G.I. 77) contain 39% sugar while Rice Krispies™ (G.I. 82) contain very little sugar. Many foods with large amounts of sugar have G.I. factors close to 60 – lower than white bread.

Sugar can be a source of enjoyment and help you limit your intake of high fat foods, but the blood sugar response to a food is hard to predict. Use the tables in this book and your own blood sugar monitoring as a guide.

WHY SHOULD YOU WATCH OUT FOR FATTY FOODS?

With diabetes, being overweight and eating fatty foods prevents insulin from doing its job. When insulin can't work properly (or there isn't enough of it) blood sugar levels rise. Most Type 2 (non-insulin dependent) diabetes is associated with an excess of abdominal fat (a 'pot belly').

Crumbed or battered foods, hot chips, fried rice, pastries or other such fatty foods are often the cause of elevated blood sugar. The high G.I. of the potato, rice or flour tends to increase blood sugar levels, and the extra fat interferes with the action of insulin and makes it less effective in clearing sugar from the blood.

Some foods high in fat have a low G.I. and may appear falsely favourable because of this. The G.I. is low because fat tends to slow the rate of stomach emptying and therefore the rate at which foods are digested in the small intestine. Some high fat foods therefore tend to have lower G.I.s than their low fat equivalents (potato crisps, 54 compared with a dry-baked potato, 85). This doesn't make them a better food.

It is important not to base your food choices solely on the G.I. factor. Low fat eating is the aim of the game not only for people with diabetes but for everyone.

DID YOU KNOW ABOUT THE SUGAR-FAT SEESAW?

Did you know that fat and sugar tend to show a reciprocal or seesaw relationship in the diet? Studies over the past decade have found that diets high in sugar are no less nutritious than low sugar diets. This is because restricting sugar is frequently followed by higher fat consumption, and most fatty foods are poor sources of nutrients.

In some cases, high sugar diets have been found to have higher micronutrient contents. This is because sugar is often used to sweeten some very nutritious foods, such as yoghurts, breakfast cereals and milk.

A low sugar (and high fat) diet has more proven disadvantages than a high sugar (and low fat) diet.

EATING THE LOW G.I. WAY

A food is not good or bad on the basis of its G.I. Eating the low G.I. way means eating a variety of foods – possibly a wider variety than you are already eating.

Usually we eat a combination of carbohydrate foods, like baked beans on toast, sandwiches and fruit, pasta and bread, cereal and toast, potatoes and corn. The G.I. of a meal consisting of a mixture of carbohydrate foods is a weighted average of the G.I. factors of the carbohydrate foods. The weighting is based on the proportion of the total carbohydrate contributed by each food. Studies show that when a food with a high G.I. factor is combined with a food with a low G.I. the complete meal has an intermediate G.I.

A rule of thumb:
High G.I. food + Low G.I. food
= Intermediate G.I. meal

As with Calories, the G.I. value is not precise. What G.I. values give you is a guide to lowering the G.I. of your day. A simple change can make a big difference. Look at the following ideas for the meals in your day and see how you could lower the G.I. factor of your diet. (See pages 30 to 49.)

ARE YOU REALLY CHOOSING LOW FAT?

There's a trick to food labels that it is worth being aware of when shopping for low fat foods. Nutrient claims are covered by food legislation that specifies what low fat really means.

Low fat means that the food must be 5% fat or less. On the nutrient table the food must not contain more than 5 g fat per 100 g food.

Reduced fat on the other hand means that the food must not contain more than 75% of the fat found in the original reference food. In other words, the amount of fat has to be reduced by at least 25% – but it isn't low enough in fat to be called a low fat food.

Where you have the choice, pick the low fat product. Better still is 'fat free' which contains less than 0.15% fat.

BREAKFAST BASICS

1. Start with some fruit or juice

Fruit contributes fibre and, more importantly, vitamin C, which helps your body absorb iron. The lowest G.I. fruits and juices are:

Cherries	22	Dried apricots	31	Grapes	46
Plums	39	Apples	38	Oranges	44
Grapefruit	25	Pears	38	Pineapple juice	46
Peaches	42	Apple juice	40	Grapefruit juice	48

2. Try some breakfast cereal

Cereals are important as a source of fibre, vitamin B and iron. When choosing processed breakfast cereals, look for those with a high fibre content. The top five low G.I. cereals are:

All-Bran™	42	Muesli	56
Sultana Bran™	52	Porridge	42
Special K™*	54		

(*Add some bran to boost its fibre content)

We know the G.I. factor for about fifteen breakfast cereals on the British market and, as more research is done, the range of low G.I. cereals will expand.

3. Add milk or yoghurt

Low fat milks and yoghurts can make a valuable contribution to your daily calcium intake by including them at breakfast. All have a low G.I. factor. Lower fat varieties have just as much, or more, calcium as full cream milk.

4. Plus some bread or toast if you like

The lowest G.I. breads are:
Mixed grain bread 47
Fruit loaf (dense type) 47

---

10 QUICK, LOW FAT, LOW G.I. BREAKFAST IDEAS

1. Spread raisin toast with low fat cream cheese and top with sliced apple.
2. Top a slice of heavy fruit loaf with sliced banana.
3. Sprinkle porridge with raisins and brown sugar.
4. Enjoy a low fat milkshake.
5. Spoon a sliced peach and raspberries through a tub of low fat yoghurt
6. Top a bowl of All-Bran™ and low fat milk with canned pear slices.

7. Smear avocado on mixed grain bread and top with baked beans.
8. Team a bowl of Sultana Bran™ and low fat milk with a glass of fresh orange juice.
9. Top a heavy grain fruit bread with fresh ricotta.
10. Enjoy a steaming hot chocolate (made with low fat milk) with wholegrain toast and Marmite.

LIGHT LUNCH IDEAS

Foods such as:

Baked beans	Mixed bean salad	Ravioli
Bread roll	Noodles	Rice salad
Chilli beans	Pasta	Steamed rice
Curried lentils	Pasta salad	Sweet corn
Fruit loaf	Pea soup	Tabbouleh
Grain bread	Pita bread	Toast
Minestrone		

1. Base your light meals on carbohydrate

2. You might add a little meat, cheese, egg, or fish

Remember – the quantity of carbohydrate should be high, the add-ons should be accents. Here are some ideas:

Bacon, just a sprinkle of lean, chopped
Cheddar cheese, a couple of small cubes

Chicken, about ¼ cup, chopped, cooked
Egg, hard boiled and quartered
Ham, a thin slice
Parmesan, a sprinkling of grated
Pastrami, a lean slice
Paté, a smearing
Roast beef, a lean slice
Sardines, a couple with lemon
Smoked oysters, 3 to 4
Smoked salmon, a slice
Tuna, a tiny tin in brine or water
Turkey breast, a thin slice
Yoghurt, a small tub of low fat

3. Fill it out with vegetables

Here are some suggestions:

Baby beets, whole
Cabbage, shredded
Carrot, grated
Cauliflower florets
Celery sticks
Cherry tomatoes
Cucumber
Spinach
Olives, a scoop
Pepper strips
Salad greens
Shallots, sliced
Peas
Sprouts
Sun-dried tomatoes

4. And round it off with fruit

10 LOW G.I. LUNCHES ON THE GO

1. Fill some flat bread with hummus and tabbouleh.
2. Top a bowl of pasta with pesto or chopped fresh herbs and ricotta.
3. Put your favourite sandwich filling on bread (toasted if you like).
4. Melt cheese over tomato on slices of grain bread.
5. Top a tub of fruit salad with a pot of low fat yoghurt.
6. Take a green salad plus some bean salad, add grainy bread.
7. A bowl of steaming home-made minestrone and a piece of fruit.
8. A lentil or vegie burger with chilli sauce and salad on a crusty roll.
9. Fill a jacket potato with baked beans and top with a sprinkle of cheese.
10. Beat up a banana smoothie and couple it with a high fibre apple muffin.

MASTERING A LOW G.I. MAIN HEAL

1. First choose the carbohydrate

Which will it be? Potato (new) or sweet potato, rice (Basmati), pasta (any type), grains (like cracked wheat or barley), chickpeas, lentils and beans or a combination? Could you add some bread or corn?

2. Add vegetables – and lots of them

Fresh, frozen or canned – the more the merrier.

Artichokes	Cauliflower	Onions
Asparagus	Celery	Peas
Beans	Chinese greens	Peppers
Broccoli	Eggplant	Silverbeet
Brussels sprouts	Fennel	Squash
Cabbage	Leeks	Witloof
Carrots	Mushrooms	Zucchini
	Okra	

3. **Now, just a little protein for flavour and texture**

Remember, you don't need much – some slivers of beef to stir-fry, a sprinkle of tasty cheese, strips of ham, a dollop of ricotta, a tender chicken breast, slices of salmon, a couple of eggs, a handful of nuts, or use the protein found in your grains and legumes.

4. Think twice about using any fat

Check that you are using a healthy type (mono- or poly-unsaturated).

10 LOW G.I. DINNER IDEAS

1. Team Spaghetti Bolognese with a green salad.
2. Wrap a fish fillet dressed with herbs and lemon, or tomato and onion, in foil and bake. Serve with a heavy grain bread roll, mixed vegetables or salad.
3. Stir-fry chicken, meat or fish with mixed green vegetables. Serve with Basmati rice or Chinese noodles.
4. Grill a steak and serve with a trio of low G.I. vegetables – new potato, sweet corn and peas.
5. Make a quick and easy spicy dhal with lentils and simply serve with rice and chutney.
6. Cook spinach and ricotta tortellini, team up with fresh garden vegetables and top with a tomato pasta sauce.
7. Create a one-pot chicken casserole with your favourite vegetables and chunks of new potatoes.
8. Make a lasagne – vegetables and beans or beef and serve it with salad.
9. Buy a barbecued chicken, steam sweet corn cobs and toss a salad together.
10. Serve chilli beans and beef with a soft tortilla bread.

Jennie's tip for those pressed for time: Buy one takeaway meal, split it with your partner and pad it out with home-cooked low G.I. rice or pasta plus loads of vegetables.

DESSERTS:A LOW G.I. FINISH

Although often overlooked, desserts can make a valuable contribution to your daily calcium and vitamin C intake when they are based on low fat dairy foods and fruits. Recipes incorporating fruit for sweetness will have more fibre and a lower G.I. than recipes with sugar. What's more, desserts are usually carbohydrate rich which means they help top-up our satiety centre, signifying the completion of eating and reducing the tendency for late night nibbles.

If you haven't time to prepare a dessert, why not simply serve a bowl of fruits in season or a fruit platter with ricotta cheese? Remember, temperate climate fruits such as apples, pears and stone fruits tend to have the lowest G.I. values.

Apples
Apricots
Bananas
Blueberries
Cherries
Custard apples
Dates
Figs
Grapefruit
Grapes
Honeydew melon
Kiwi fruit
Lychees
Mandarins
Mangos
Nectarines
Oranges
Pawpaw
Peaches
Pears
Persimmons
Pineapple
Plums
Prunes
Quinces
Raspberries
Rhubarb
Star fruit
Strawberries
Tamarillo
Watermelon

10 QUICK AND EASY LOW G.I. DESSERTS

1. Low fat ice cream and strawberries.
2. Baked whole apple, stuffed with dried fruit.
3. Fruit salad with low fat yoghurt.
4. Make a fruit crumble – top cooked fruit with a crumbled mixture of toasted muesli, wheat flakes, a little melted margarine (poly- or mono-unsaturated) and honey.
5. Slice a firm banana into some low fat custard.
6. Top canned fruit (peaches or pears) with low fat ice cream or low fat custard.
7. Wrap pie-packed apple, sultanas, currants and spice, in a sheet of filo pastry (brushed with milk, not fat) and bake as a strudel.
8. Make a winter fruit salad with segments of citrus fruits plus raisins soaked in orange juice, honey and brandy.
9. Team canned plums with a dollop of natural yoghurt and a sprinkle of toasted muesli.
10. Enjoy a jelly whip made with low fat milk or yoghurt and set with fruit

BETWEEN-MEAL SNACKS

Many people with diabetes need between-meal snacks. The G.I. factor is especially important when carbohydrate is eaten by itself and not as part of a mixed meal. This is because carbohydrate tends to have a stronger effect on our blood sugar level when it is eaten alone.

When choosing a between-meal snack, pick a low fat one with a low G.I. factor. For example, an apple with a G.I. of 36 is better than a slice of white bread with a G.I. of around 70, and will result in less of a jump in the blood sugar level.

Some snack foods with a very low G.I. (such as peanuts, 14) have a very high fat content and are not recommended for people with a weight problem. As an occasional snack they are fine, especially as their fat is monounsaturated, but not every day. Remember, peanuts are very more-ish and it is often hard to stop at just a handful!

New evidence suggests that the people who graze, eating small amounts of food throughout the day at frequent intervals, may actually be doing themselves a favour. A recent study showed that snacking stimulates the body to use up more energy for metabolism compared to concentrating the same amount of food into three meals.

SUSTAINING SNACKS

- An apple
- An apple and oatbran muffin
- Dried apricots
- A mini can of baked beans
- A small bowl of cherries
- Ice cream (low fat) in a cone
- Milk, milkshake or smoothie (low fat, of course)
- Oatmeal biscuits, Highland, (2 to 3)
- An orange
- 200 ml orange juice, freshly squeezed
- Pitta bread and Marmite
- A big bowl of popcorn, low fat of course
- 1–2 slices raisin toast – try the grain-based fruit loaves
- Grain bread sandwich with your favourite filling
- Sultana Bran™, a bowl with low fat milk
- Sultanas, a small box
- 200 g tub of low fat yoghurt

A WEEK OF LOW G.I. EATING

This week of menus shows how to achieve a healthy diet with a low G.I. Use the menus for ideas of foods to include for low fat, high carbohydrate, low G.I. eating every day.

We have included between-meal snacks in most of the menus as people who take tablets or insulin to help control their diabetes generally need them. While not everyone with diabetes has to eat between meals, snacks can be a normal part of a healthy diet.

We have analysed each menu to estimate its Calories and the amount of fat, carbohydrate and fibre. We have also calculated the G.I. factor. By emphasising low G.I. foods, we have created menus with a predicted low G.I., that is, less than 55. Here are some points to consider.

1. Calories

Calories are a measure of the total amount of energy available from a food. Each day you require a certain amount of Calories to fuel your body. The younger and more active you are the more Calories you need. If you want to lose weight you may need to reduce your Calorie intake. The Calorie levels of these menus are reasonably low and represent a minimum requirement for most people.

2. Fat

People with diabetes are advised to eat a low fat diet. The menus are examples of low fat meals if you use skimmed milk and minimal amounts of margarine and other fats and oils when preparing them. If you are trying to lose weight, aim for a fat intake of 30 to 50 g per day. Remember, such low fat diets are not recommended for young children.

3. Carbohydrate

These menus are all high in carbohydrate, deriving about half the total Calories from carbohydrate. For a small eater this is around 150 to 200 g of carbohydrate per day. The more active you are the more carbohydrate you need. Note that 150 to 200 g carbohydrate equates to 10 to 15 carbohydrate exchanges (for people who use that system).

4. Fibre

Guidelines suggest you need at least 30 g of fibre each day. This is often difficult to achieve on a low Calorie intake without eating a high fibre cereal each day. The fibre content of these menus varies from 20 to 40 g per day, giving a daily average of 30 g.

MONDAY MENU

G.I. Factor:	*43*
Total Energy:	*1640 kcal*
Fat:	*30 g*
Carbohydrate:	*227 g*
Fibre:	*42 g*

Breakfast	Cereal and toast Weetabix™, 2 biscuits, 30 g with ¾ cup low fat milk; 1 slice mixed grain loaf toasted with a scrape of margarine and a dollop of honey
Mid-morning	An apple
Lunch	Tuna-topped open sandwich and salad Combine about 60 g canned tuna with a little shallot, parsley and low fat mayonnaise. Pile it on top of sliced tomato on a slice of mixed grain bread and team with 220 g mixed bean salad and salad greens
Mid-afternoon	2 cups of popped corn
Dinner	Char-grilled steak and vegetables Grill or barbecue a lean piece of steak (about 120 g raw weight), basting with marinade if desired. Cook a new potato in its skin and serve with a small cob of corn, baby carrots and green peas.
Later	A low fat fruit yoghurt

TUESDAY MENU

G.I. Factor:	*50*
Total Energy:	*1555 kcal*
Fat:	*40 g*
Carbohydrate:	*220 g*
Fibre:	*20 g*

Breakfast	Porridge Cook ½ cup rolled oats with ½ cup water and ½ cup low fat milk. Serve with a teaspoon of brown sugar or a tablespoon of sultanas
Mid-morning	An orange
Lunch	A sandwich on the run Cheese and salad on wholemeal bread
Mid-afternoon	A 'diet' yoghurt
Dinner	Pasta and sauce Top about 1½ cups cooked pasta with tomato or Bolognese sauce and a sprinkle of grated Parmesan cheese. Include a small fresh bread roll and a glass of red wine if desired.
Later	A bunch of grapes

WEDNESDAY MENU

G.I. Factor:	*40*
Total Energy:	*1460 kcal*
Fat:	*43 g*
Carbohydrate:	*207 g*
Fibre:	*36 g*

Breakfast	Cereal, fruit and toast ½ cup All-Bran with ½ cup canned peaches and ¾ cup low fat Milk, 1 slice of mixed grain toast with a smear of peanut butter
Mid-morning	2 oatmeal biscuits
Lunch	A toasted sandwich Ham and tomato toasted sandwich on mixed grain loaf. Finish with an apple
Mid-afternoon	An orange
Dinner	Stir-fry and rice Stir-fry lean beef strips (about 100 g raw weight) and a combination of vegetables (like broccoli, zucchini, shallots, cabbage), adding soy sauce, ginger, garlic etc. to flavour. Serve with a cup of boiled rice (Basmati)
Later	A juicy peach or other fresh fruit in season

THURSDAY MENU

G.I. Factor:	*48*
Total Energy:	*1480 kcal*
Fat:	*24 g*
Carbohydrate:	*230 g*
Fibre:	*38 g*

Breakfast	Muesli, fruit and yogurt ½ cup Swiss style muesli with 150 g low fat fruit yoghurt and a fresh chopped pear
Mid-morning	2 Ryvita with Marmite
Lunch	A filled spud Microwave a large potato in its jacket. Slice off the top, scoop out the middle mixing the potato with 1 tablespoon each of cottage cheese, diced ham, reduced fat Cheddar cheese and a sprinkle of chopped shallots. Pile this mixture into the potato cup and reheat
Mid-afternoon	200 ml (¾ cup) low fat flavoured milk
Dinner	Mexican roll-ups Fill a round of pitta bread with chilli beans (120 g kidney beans) and beef (80 g), and shredded lettuce, diced tomato and a sprinkle of reduced fat Cheddar cheese
Later	An apple

FRIDAY MENU

G.I. Factor:	*44*
Total Energy:	*1695 kcal*
Fat:	*50 g*
Carbohydrate:	*210 g*
Fibre:	*30 g*

Breakfast	Egg, toast and orange juice Boil an egg, lop the top off it and enjoy with whole grain toast fingers and Marmite Have another piece of toast with a smear of margarine and a dollop of marmalade
Mid-morning	Small glass of orange juice
Lunch	A healthy kebab Tabbouleh, felafel, lettuce, tomato, cheese all wrapped in pitta bread
Mid-afternoon	A couple of kiwi fruit
Dinner	Pasta with chicken and mushroom sauce Make up a light white sauce base, add some sautéed mushrooms, seasonings and cooked, chopped chicken and serve over steaming pasta (about 1 cup)
Later	A slice of raisin toast with light cream cheese

SATURDAY MENU

G.I. Factor:	*42*
Total Energy:	*1555 kcal*
Fat:	*25 g*
Carbohydrate:	*225 g*
Fibre:	*30 g*

Breakfast	Banana smoothie and toast Whip up a banana, low fat yoghurt, milk and a dollop of honey in a blender to make a quick liquid breakfast. Serve with a slice of toast and Marmite if you still feel hungry, or keep it for mid-morning
Lunch	Cheese and tomato melts Top 2 pieces of wholegrain bread with sliced tomato and a low fat cheese slice. Heat under a grill until the cheese melts. Finish off with an orange
Mid-afternoon	A crisp apple
Dinner	Roast pork with apple sauce Serve lean roast pork (about 120 g) with a dollop of apple sauce, dry roasted potato, pumpkin and greens. Add a glass of wine if you like and top the meal off with ½ a mango, and strawberry low fat yoghurt

SUNDAY MENU

G.I. Factor:	*50*
Total Energy:	*1390 kcal*
Fat:	*28 g*
Carbohydrate:	*200 g*
Fibre:	*25 g*

Breakfast	Raisin toast and hot chocolate. Spread 2 slices of toasted Fruit Loaf with ricotta cheese. Serve with hot chocolate made with low fat milk
Mid-morning	A muesli bar on the go
Lunch	A toasted baked bean sandwich Spread some baked beans on a couple of slices of mixed grain loaf and pop in the toasted sandwich maker
Mid-afternoon	An ice cream in a cone Low fat ice creams or frozen yoghurt
Dinner	Fish and chips and salad. Cut a new potato into chips, spray with cooking oil. spread on baking tray and bake in a very hot oven until browned. Wrap a piece of fresh fish (about 155 g) in foil, season and add lemon. Bake it for the last 10 to 15 minutes with the chips. Toss together lettuce, cucumber and shallots with vinegar and drizzle of olive oil
Later	A bunch of grapes or other fruit

5 LITTLE TIPS THAT MAKE A BIG DIFFERENCE

- Think of carbohydrate foods as the number one priority in your meals.
- Change a staple in your diet, like bread, to a low G.I. type to make a big difference to the G.I. factor of your day.
- Get in touch with your true appetite and use it to guide the amount of food you eat. Low fat, high fibre, low G.I. factor foods fill you up best.
- Try to eat at least two low G.I. factor meals each day.
- Mix high G.I. foods with low G.I. foods in your meals – the combination will give an overall intermediate G.I.

Did you know?

Eating more Calories than your body needs is less likely to occur if the foods eaten are carbohydrate rich and have a low G.I. factor. Using these foods, you will feel full (or even over full) and satisfied before you have overstepped your body's energy needs.

CHECKLIST OF LOW G.I. FOODS

Here is a checklist of foods with the lowest G.I. factors. The G.I. factor is given next to each food (based on glucose = 100).

* Foods containing high levels of fat.

BREAKFAST CEREALS

Rice bran	19
All-Bran (Kellogg)	42
Porridge	42
Sultana Bran™ (Kellogg)	52
Special K™ (Kellogg)	54
Oat bran (unprocessed)	55
Muesli	56

BREADS

Kibbled wheat loaf (made with 50% grain)	43
Fruit Loaf	47
Oat bran bread (made with 50% bran)	47
Pumpernickel	50

VEGETABLES

Peas	48
Sweetcorn	55
Yam	51
Sweet potato	54

BISCUITS AND CAKES

Apple muffin*	44
Sponge cake	46
Banana cake*	47
Highland Oatmeal	55
Rich Tea	55

JUICES

Apple juice (unsweetened)	40
Orange juice	46
Pineapple juice (unsweetened)	46
Grapefruit juice	48

SNACK FOODS

Chocolate*	49
Potato crisps*	54
Popcorn	55

CEREAL GRAINS AND PASTA

Pearled barley	25
Spaghetti (average)	41
Macaroni	45
Instant noodles	47
Bulgur (cracked wheat)	48
Buckwheat	49
High amylose rice (Basmati)	54
Quick Cooking Wheat	54

DAIRY FOODS

Yoghurt (low fat, artificially sweetened)	14
Milk	27
Skimmed milk	32
Yoghurt (low fat, fruit and sugar sweetened)	33
Chocolate flavoured milk	34
Custard	43
Ice cream (low fat)	50

LEGUMES		FRUIT	
Soya beans (canned)	14	Cherries	22
Soya beans	18	Grapefruit	25
Lentils (red)	26	Dried apricots	31
Kidney beans	27	Pears	38
Lentils (green)	30	Apples	38
Butter beans	31	Plums	39
Lima beans	32	Peaches	42
Split peas	32	Oranges	44
Chick peas	33	Grapes	46
Haricot/Navy beans	38	Canned peaches	47
Chick peas (canned)	42	Kiwi fruit	52
Baked beans (canned)	48		
Kidney beans (canned)	52		

Note: Canned legumes have higher G.I.s than the boiled varieties because the temperatures and pressures used in the canning process increase the digestibility of the starch. But, canned legumes are still an excellent low fat, high fibre, nutrient-rich low G.I. choice!

SUBSTITUTING LOW G.I. FOR HIGH G.I. FOODS

High G.I. Food	*Low G.I. Alternative*
Bread, wholemeal or white	Bread containing a lot of wholegrains
Processed breakfast cereal	Unrefined cereal such as oats or check the G.I. list for processed cereals with a low G.I. factor e.g. All-Bran™
Plain biscuits and crackers	Biscuits made with dried fruit and wholegrains such as oats
Cakes and muffins	Look for those made with fruit, oats, wholegrains
Tropical fruits such as bananas	Temperate climate fruits such as apples and stone fruit
Potato	Use new potatoes, sweet potatoes, pasta, noodles, butter beans
Rice	Try Basmati

THE LOW G.I. PANTRY

To make the low G.I. choices, easy choices, you need to keep the right foods in your pantry and refrigerator.

Breads

Mixed grain loaves with cracked grains
Fruit loaf (the heavy types. Keep a loaf in the freezer)

Cereals

All-Bran™ (Kellogg)
Sultana Bran™ (Kellogg)
Muesli
Rolled oats
Oat bran
Rice bran

Pasta and rice

Pearl barley
Basmati rice
Pasta of various shapes and flavours

Pulses

Canned or dried lentils (red and brown), legumes (chick peas, cannellini beans)
A variety of canned legumes (kidney beans, mixed beans, baked beans)

Vegetables

Canned sweet corn

Other canned vegetables like canned tomatoes, asparagus, peas, mushrooms are always handy to boost the vegetable content of a meal

Canned crushed tomatoes and tomato puree, bottled tomato

Sauces and dressings

Pasta sauces

Prepared chicken stock or stock cubes

Low oil salad dressings

Tomato paste

Fruits

Dried fruits – sultanas, dried apricots, fruit medley, raisins, prunes etc.

Canned peaches, pears, apple

Milk products

Long-life skimmed milk or skimmed milk powder

Canned evaporated skimmed milk

Custard powder

Flavourings

Spices – curry powder, cumin, turmeric, mustard etc.

Herbs – oregano, basil, thyme etc.

Bottled minced ginger, chilli and garlic

WHAT TO KEEP IN THE REFRIGERATOR AND FREEZER

Dairy foods

Skimmed or fat-reduced milk

Low fat natural yoghurt

Low fat fruit yoghurt

Low fat ice cream

Frozen low fat yoghurt, sorbet

Eggs

Cheese

- low fat processed slices
- fat-reduced or low fat Cheddar
- grated Parmesan
- cottage or ricotta cheese

Frozen peas, sweet corn, spinach, carrots etc.

Frozen berries

HYPOGLYCAEMIA – THE EXCEPTION TO THE LOW G.I. RULE

If you take insulin or tablets to treat diabetes, your blood sugar level may sometimes fall below 4 millimoles per litre, which is the lower end of the normal range. When this happens you might feel hungry, shaky, sweaty and be unable to think clearly. This is called a hypo (short for 'hypoglycaemia').

A hypo is a potentially dangerous situation and must be treated straight away by eating some carbohydrate food. In this case you should pick a carbohydrate with a high G.I. factor because you need to increase your blood sugar quickly. Jelly beans (with a G.I. factor of 80) are a good choice. If you are not due for your next meal or snack you should also have some low G.I. carbohydrate, like an apple, to keep your blood sugar from falling again until you next eat.

Glucose tablets or jelly beans are good choices for treating a 'hypo'.

DON'T FORGET ABOUT EXERCISE

Diet is by no means all there is to managing diabetes. Diabetes is with you for the rest of your life so looking after yourself requires healthy lifestyle habits.

These days it is increasingly easy to overeat. Refined foods, convenience foods and fast foods frequently lack fibre and conceal fat so that before we feel full, we have overdosed on Calories.

It is even easier not to exercise. With intake exceeding output on a regular basis, the result for too many of us is a lounge lizard lifestyle and subsequent weight gain.

WHY EXERCISE KEEPS YOU MOVING

The effect of exercise doesn't stop when you stop moving. People who exercise have higher metabolic rates and their bodies burn more Calories per minute even when they are asleep!

HOW EXERCISE HELPS

Regular physical activity can reduce our blood sugar levels, lower our risk of heart and blood vessel disease, lower high blood pressure, increase stamina, reduce stress and help us relax. It is meant for us all.

- Exercise speeds up our metabolic rate. By increasing our Calorie expenditure, exercise helps to balance our sometimes excessive Calorie intake from food and helps us control our weight.
- Exercise makes our muscles better at using fat as a source of fuel. By improving the way insulin works, exercise increases the amount of fat we burn.

A low G.I. diet has the same effect. Low G.I. foods reduce the amount of insulin we need which makes fat easier to burn and harder to store. Since body fat is what you want to get rid of when you lose weight, exercise in combination with a low G.I. diet makes a lot of sense!

Remember that reduction in body weight takes time. Even after you've made changes in your exercise habits your weight may not be any different on the scales. This is particularly true in women whose bodies tend to adapt to increased Calorie expenditure.

SO WHAT CAN YOU DO?

We need to adapt our lifestyle to our more Calorie-laden diet and fewer physical demands. It's important to catch bursts of physical activity wherever we can to increase our energy output.

Your increased movement may be planned, for example walking 10 to 15 minutes, 5 to 6 days a week. Home-based walking programs like this seem to be one of the best strategies for increasing physical output.

Besides planned activity you simply need to move around more in your day. Take a 10 minute walk at lunchtime, walk to the shops to get the Sunday paper, park a kilometre from work, or take the dog for a walk each night. Try one car-free and one TV-free day each week.

Personal trainers can improve adherance to an exercise program, if you are willing to pay for the privilege of being regularly motivated by someone with appropriate skills.

Whatever it takes for you, do it. Regard movement as an opportunity to improve your physical well-being and not an inconvenience.

HOW TO MAKE YOUR EXERCISE SUCCESSFUL

Eight key factors to make exercise successful include:

- seeing a benefit for yourself
- enjoying what you do
- feeling that you can do it fairly well
- fitting in with your daily life
- being inexpensive
- being accessible
- being safe
- being socially acceptable to your peers

USING THE G.I. FACTOR WHEN YOU EXERCISE

Here we're talking about the everyday sort of moderate exercise that all of us should be doing. If you train physically hard a number of days a week and perhaps compete in sports you should read *The Pocket Guide to the Glucose Revolution and Sports Nutrition.*

It is sometimes necessary with diabetes to eat extra carbohydrate when you exercise. This depends on the type of diabetes you have and the type and amount of medication you take. Often, you won't want to increase your food intake – because the exercise is intended to burn off some earlier overconsumption! (For people with insulin-dependent diabetes remember this will only work if you have enough insulin in your body and your blood sugars aren't overly high to start with.)

You may need extra carbohydrate before you exercise, or, if the exercise is prolonged over an hour or more, you may need extra carbohydrate during it too. Whether or not you need to eat extra and how much to take depends on your blood sugar level before, during and after the exercise and how your body responds to the exercise—all of which you learn from experience. Discuss your situation and how it is best to manage it with a dietitian, diabetes educator or doctor.

If you need to eat immediately before exercise to bring your blood sugar up, it makes sense to eat some high G.I. carbohydrate, such as a slice of regular bread, a couple of Morning Coffee biscuits or a ripe banana.

If you plan to eat your last meal or snack 1 to 2 hours before your exercise, it makes sense to make this a low G.I. meal to sustain you through the exercise – a sandwich made with low G.I. bread, a tub of yoghurt, baked beans on low G.I. toast, an apple.

If you need to eat something quickly after or during exercise to restore your blood sugar level, use high G.I. food – crispbread, a bowl of Cornflakes or Rice Krispies, a slice of watermelon.

Nothing can replace the benefit of measuring your blood sugar when you exercise to assess your body's response and judge your carbohydrate needs.

UNDERSTANDING THE G.I. FACTOR

The glycaemic index concept was first developed by Dr David Jenkins, a professor of nutrition at the University of Toronto, Canada, to help determine which foods were best for people with diabetes. At that time, the diet for people with diabetes was based on a system which had a less strong scientific basis.

The carbohydrate exchange system assumed that all starchy foods produce the same effect on blood sugar levels, even though earlier studies had already proven this was not correct. Jenkins was one of the first people to question this assumption and investigate how real foods really behave in the bodies of real people.

Since then, scientists, including the authors of this book, have tested the effect of many foods on blood sugar levels and clinical studies in the United Kingdom, France, Italy, Australia and Canada all have proven without doubt the value of the glycaemic index.

The glycaemic index (or G.I. factor) of foods is simply a ranking of foods based on their immediate effect on blood sugar levels. To make a fair comparison, all foods are compared with a reference food such as pure glucose and are tested in equivalent carbohydrate amounts.

THE KEY IS THE RATE OF DIGESTION

Carbohydrate foods that break down quickly during digestion have the highest G.I. factors. Conversely, carbohydrates which break down slowly, releasing glucose gradually into the bloodstream have low G.I. factors.

For most people most of the time, the foods with low G.I.s have advantages over those with high G.I. values.

The higher the G.I., the higher the blood sugar levels after consumption of the food. Rice Krispies™ (G.I. 82) and baked potatoes (G.I. 85) have very high G.I. factors, meaning their effect on blood sugar levels is almost as high as that of an equal amount of pure glucose (yes, you read it correctly).

Low G.I., less than 55
Intermediate G.I., 55 to 70
High G.I., more than 70

Figure 1 shows the blood sugar response to potatoes compared with pure glucose. Foods with a low G.I. (like lentils at 28) show a flatter blood sugar response when eaten, as shown in Figure 2. The peak blood sugar level is lower and the return to baseline levels is slower than with a high G.I. food.

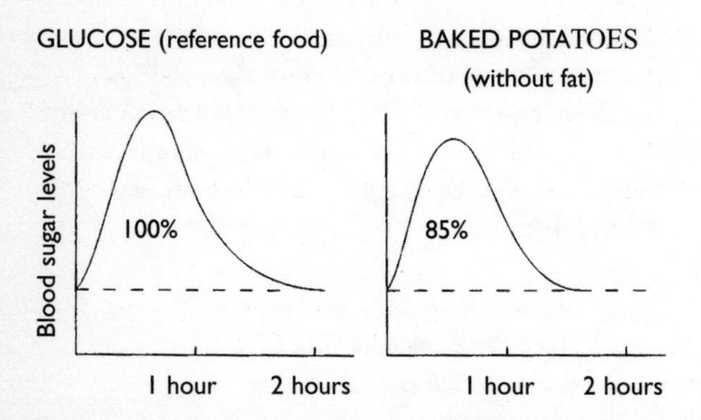

Figure 1. The effect of pure glucose (50 g) and baked potatoes without fat (50 g carbohydrate portion) on blood sugar levels.

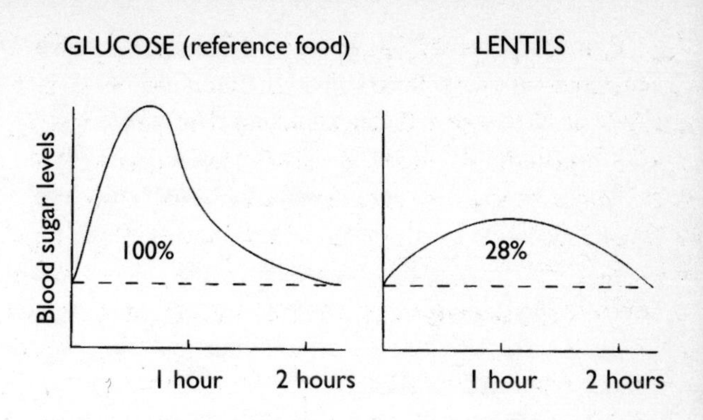

Figure 2. The effect of pure glucose (50 g) and lentils (50 g carbohydrate portion) on blood sugar levels.

IF A FOOD HAS A HIGH G.I., SHOULD SOMEONE WITH DIABETES AVOID IT.

Some foods like bread and potatoes have high G.I.s (70 to 80). But, potatoes and bread can play a major role in a high carbohydrate and low fat diet. You only have to exchange about half the carbohydrate (from high to low G.I.) to achieve improvements. So, there's plenty of room for bread and potatoes. Some breads have a lower G.I. than others. Choose these ones if your goal is to lower the G.I. as much as possible.

You can't predict the G.I. of a food from its composition. To test the G.I., you need real people and real foods. We describe how the G.I. factor is measured in the following section. Standardised methods are always followed so that results from one group of people can be directly compared with those from another.

WHAT GIVES ONE FOOD A HIGH G.I. AND ANOTHER A LOW ONE?

The physical state of the starch in the food is the most important factor influencing the G.I. That's why food processing has such a profound effect on the G.I. factor.

5 KEY FACTORS THAT INFLUENCE THE G.I.

Cooking methods

Cooking and processing increases the G.I. of a food because it increases the amount of gelatinised starch in the food (e.g. cornflakes).

Physical form of the food

An intact fibrous coat (e.g. in grains and legumes) acts as a physical barrier and slows down digestion, lowering the G.I. factor.

Type of starch

There are two types of starch in foods, amylose and amylopectin. The more amylose starch a food contains, the lower the G.I. factor.

Fibre

Viscous, soluble fibres, like those found in rolled oats and apples, slow down digestion and lower a food's G.I. factor.

Sugar

The presence of sugar and the type of sugar will influence the G.I. Fruits such as apples and oranges with a low G.I. are high in fructose.

HOW WE MEASURE THE G.I. FACTOR

Pure glucose produces the greatest rise in blood sugar levels. All other foods have less effect when fed in equal carbohydrate quantities. The G.I. of pure glucose is set at 100 and every other food is ranked on a scale from 0 to 100 according to its actual effect on blood sugar levels.

1. To find out the G.I. of a food, a volunteer eats an amount of that test food containing 50 g of carbohydrate (calculated from food composition tables) – 50 g of carbohydrate is equivalent to 3 tablespoons of pure glucose powder.
2. Over the next 2 hours (3 hours if the volunteer has diabetes), we take a blood sample every 15 minutes during the first hour and every 30 minutes thereafter. We measure and record the blood sugar level of these samples.
3. The blood sugar level is plotted on a graph and the area under the curve is calculated using a computer program (see Figure 3).
4. We compare the volunteer's response to the test food with his/her other blood sugar response to 50 g of pure glucose (the reference food).
5. The reference food is tested on 2 or 3 separate occasions and we calculate an average value to reduce the effect of day-to-day variation in blood sugar responses.

Note: The G.I. factor of the test food is the average value of a group of 8 to 12 volunteers. Results obtained in a group of people with diabetes are comparable to those without diabetes. We refer to all foods according to a standard where glucose equals 100.

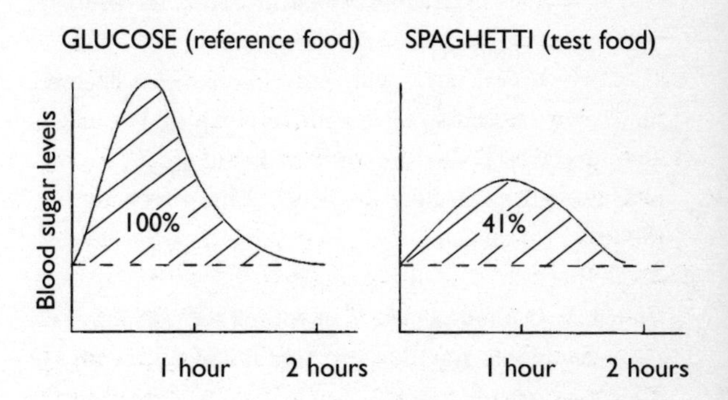

Figure 3. The effect of a food on blood sugar levels is calculated using the area under the curve (hatched area). The area under the curve after consumption of the test food is compared with the same area after the reference food (usually 50 g of pure glucose or a 50 g carbohydrate portion of white bread).

HOW TO USE THE G.I. TABLES

These simplified tables are an A to Z listing of the G.I. factor of foods commonly eaten in the British Isles. Approximately 300 different foods are listed. They include some new values for foods tested only recently.

The G.I. value shown next to each food is the average for that food using glucose as the standard, i.e., glucose has a G.I. value of 100, with other foods rated accordingly. The average may represent the mean of 10 studies of that food world wide or only 2 to 4 studies.

We have included some foods in the list which are not commonly eaten and other foods which may be encountered on overseas trips.

To check on a food's G.I., simply look for it by name in the alphabetic list. You may also find it under a food type – fruit, biscuits.

Included in the tables is the carbohydrate (CHO) and fat content of a sample serving of the food. This is to help you keep track of the amount of fat and carbohydrate in your diet. Refer to pages 19 and 26 for advice on how much carbohydrate and fat is recommended.

Remember when you are choosing foods, the G.I. factor isn't the only thing to consider. In terms of your blood sugar levels you should also consider the amount of carbohydrate you are eating. For your overall health the fat, fibre and micronutrient content of your diet is also important. A dietitian can guide you further with good food choices.

If you can't find a G.I. value for a food you eat on many occasions please email us and we'll give you an estimated value of the food and endeavour to test its G.I. in the future. Address your message to:

j.brandmiller@biochem.usyd.edu.au

The G.I. values in these tables are correct at the time of publication. However, the formulation of some commercial foods can change and the G.I. may be altered. Check our web page for revised and new data.

www.biochem.usyd.edu.au/~jennie/GI/glycemic_index.html

A-Z OF FOODS WITH G.I. FACTOR, PLUS CARBOHYDRATE & FAT COUNTER

Food	G.I.	Fat	CHO
		(grams per serving)	
All Bran™, 40 g	42	1	22
Angel food cake, 30 g	67	trace	17
Apple, 1 medium, 150 g	38	0	18
Apple juice, unsweetened, 250 ml	40	0	33
Apple muffin, 1, 80 g	44	10	44
Apricots, fresh, 3 medium, 100 g	57	0	7
canned, light syrup, 125 g	64	0	13
dried, 5–6 pieces, 30 g	31	0	13
Bagel, 1 white, 70 g	72	1	35
Baked beans, canned in tomato sauce, 120 g	48	1	13
Banana cake, 1 slice, 80 g	47	7	46
Banana, raw, 1 medium, 150 g	55	0	32
Barley, pearled, boiled, 80 g	25	1	17
Basmati white rice, boiled, 180 g	58	0	50
Beetroot, canned, drained, 2–3 slices, 60 g	64	0	5
Bengal gram dhal, 100 g	54	5	57
Biscuits			
Digestives, plain, 2 biscuits, 30 g	59	6	21

Food	G.I.	Fat	CHO
		(grams per serving)	
Biscuits (*continued*)			
Milk Arrowroot, 2 biscuits, 16 g	63	2	13
Morning Coffee, 3 biscuits, 18 g	79	2	14
Oatmeal, 3 biscuits, 30 g	54	6	19
Rich Tea, 2 biscuits, 20 g	55	3	16
Shortbread, 2 biscuits, 30 g	64	8	19
Vanilla wafers, 6 biscuits, 30 g	77	5	21
Wheatmeal, 2 biscuits, 16 g	62	2	12
see also Crackers			
Black bean soup, 220 ml	64	2	82
Black beans, boiled, 120 g	30	1	26
Black gram, soaked and boiled, 120 g	43	1	16
Blackbread, dark rye, 1 slice, 50 g	76	1	21
Blackeyed beans, soaked, boiled, 120 g	42	1	24
Blueberry muffin, 1, 80 g	59	8	41
Bran			
Oat bran, 1 tablespoon, 10 g	55	1	7
Rice bran, extruded, 1 tablespoon, 10 g	19	2	3
Bran Buds™, breakfast cereal, 30 g	58	1	14

Food	G.I.	Fat	CHO
		(grams per serving)	
Bran muffin, 1, 80 g	60	8	34
Breads			
Dark rye, Blackbread, 1 slice, 50 g	76	1	21
Dark rye, Schinkenbröt, 1 slice, 50 g	86	1	22
French baguette, 30 g	95	1	15
Fruit loaf, heavy, 1 slice, 35 g	47	1	18
Gluten-free bread, 1 slice, 30 g	90	1	14
Hamburger bun, 1 prepacked bun, 50 g	61	3	24
Light rye, 1 slice, 50 g	68	1	23
Linseed rye, 1 slice, 50 g	55	5	21
Melba toast, 4 squares, 30 g	70	1	19
Pitta bread, 1 piece, 65 g	57	1	38
Pumpernickel, 2 slices	41	2	35
Rye bread, 1 slice, 50 g	65	1	23
Sourdough rye, 1 slice, 50 g	57	2	23
Vogel's™, Honey & Oat loaf, 1 slice, 40 g	55	3	17
White (wheat flour), 1 slice, 30 g	70	1	15
Wholemeal (wheat flour), 1 slice, 35 g	69	1	14

Food	G.I.	Fat	CHO
		(grams per serving)	
Bread stuffing, 60 g	74	5	17
Breadfruit, 120 g	68	1	17
Breakfast cereals			
All-Bran™, 40 g	42	1	22
Bran Buds™, 30 g	58	1	14
Cheerios™, 30 g	74	2	20
Coco Pops™, 30 g	77	0	26
Cornflakes, 30 g	84	0	26
Mini Wheats™ (whole wheat), 30 g	58	0	21
Muesli, toasted, 60 g	43	9	33
Muesli, non-toasted, 60 g	56	6	32
Oat bran, raw, 1 tablespoon, 10 g	55	1	7
Porridge (cooked with water), 245 g	42	2	24
Puffed wheat, 30 g	80	1	22
Rice bran, 1 tablespoon, 10 g	19	2	3
Rice Krispies™, 30 g	82	0	27
Shredded wheat, 25 g	67	0	18
Special K™, 30 g	54	0	21
Sultana Bran™, 45 g	52	1	35
Sustain™, 30 g	68	1	25

Food	G.I.	Fat	CHO
		(grams per serving)	
Breakfast cereals (*continued*)			
Weetabix™, 2 biscuits, 30 g	69	1	19
Broad beans, frozen, boiled, 80 g	79	1	9
Buckwheat, cooked, 80 g	54	3	57
Bun, hamburger, 1 prepacked bun, 50 g	61	3	24
Burghul, cooked, 120 g	48	0	22
Butter beans, boiled, 70 g	31	0	13
Cakes			
Angel food cake, 1 slice, 30 g	67	trace	17
Banana cake, 1 slice, 80 g	47	7	46
Flan, 1 slice, 80 g	65	5	55
Pound cake, 1 slice, 80 g	54	15	42
Sponge cake, 1 slice, 60 g	46	16	32
Cantaloupe melon, raw, ¼ small, 200 g	65	0	6
Capellini pasta, boiled, 180 g	45	0	53
Carrots, peeled, boiled, 70 g	49	0	3
Cereal grains			
Barley, pearled, boiled, 80 g	25	1	17
Buckwheat, cooked, 80 g	54	3	57
Burghul, cooked, 120 g	48	0	22
Couscous, cooked, 120 g	65	0	28

Food	G.I.	Fat	CHO
		(grams per serving)	
Cereal grains (*continued*)			
Maize			
Cornmeal, wholegrain, cooked, 40 g	68	1	30
Sweet corn, canned, drained, 80 g	55	1	16
Taco shells, 2 shells, 26 g	68	6	16
Millet Ragi, cooked, 120 g	71	0	12
Rice			
Basmati, white, boiled, 180 g	58	0	50
Tapioca (boiled with milk), 250 g	81	10.5	51
Cheerios™, breakfast cereal, 30 g	74	2	20
Cherries, 20, 80 g	22	0	10
Chick peas, canned, drained, 95 g	42	2	15
Chick peas, boiled, 120 g	33	3	22
Chocolate, milk, 6 squares, 30 g	49	8	19
Coco Pops™, breakfast cereal, 30 g	77	0	26
Condensed milk, sweetened, ½ cup, 163 g	61	15	90
Corn bran, breakfast cereal, 30 g	75	1	20
Corn chips, Doritos™ original, 50 g	42	11	33
Cornflakes, breakfast cereal, 30 g	84	0	26

Food	G.I.	Fat	CHO
		(grams per serving)	
Cornmeal (maizemeal), cooked, 40 g	68	1	30
Couscous, cooked, 120 g	65	0	28
Crackers			
Premium soda crackers, 3 biscuits, 25 g	74	4	17
Puffed crispbread, 4 biscuits, wholemeal, 20 g	81	1	15
Rice cakes, 2 cakes, 25 g	82	1	21
Ryvita™, 2 slices, 20 g	69	1	16
Stoned wheat thins, 5 biscuits, 25 g	67	2	17
Water biscuits, 5, 25 g	78	2	18
Croissant, 1	67	14	27
Crumpet, 1, toasted, 50 g	69	0	22
Custard, 175 g	43	5	24
Dairy foods			
Ice cream, full fat, 2 scoops, 50 g	61	6	10
Ice cream, low fat, 2 scoops, 50 g	50	2	13
Milk, full fat, 250 ml	27	10	12
Milk, skimmed, 250 ml	32	0	13

Food	G.I.	Fat	CHO
		(grams per serving)	
Dairy foods (*continued*)			
Milk, chocolate flavoured, low-fat, 250 ml	34	3	23
Custard, 175 g	43	5	24
Yoghurt			
low-fat, fruit, 200 g	33	0	26
low-fat, artificial sweetener, 200 g	14	0	12
Dark rye bread, Blackbread, 1 slice, 50 g	76	1	21
Dark rye bread, Schinkenbröt, 1 slice, 50 g	86	1	22
Digestive biscuits, 2 plain, 30 g	59	6	21
Doughnut with cinnamon and sugar, 40 g	76	8	16
Fanta™, soft drink, 1 can, 375 ml	68	0	51
Fettucini, cooked, 180 g	32	1	57
Fish fingers, oven-cooked, 5 x 25 g fingers, 125 g	38	14	24
Flan cake, 1 slice, 80 g	65	5	55
French baguette bread, 30 g	95	1	15
French fries, fine cut, small serving, 120 g	75	26	49

Food	G.I.	Fat	CHO
		(grams per serving)	
Fructose, pure, 10 g	23	0	10
Fruit cocktail, canned in natural juice, 125 g	55	0	15
Fruit loaf, heavy, 1 slice, 35 g	47	1	18
Fruits and fruit products			
Apple, 1 medium, 150 g	38	0	18
Apple juice, unsweetened, 250 ml	40	0	33
Apricots, fresh, 3 medium, 100 g	57	0	7
canned, light syrup, 125 g	64	0	13
dried, 5–6 pieces, 30 g	31	0	13
Banana, raw, 1 medium, 150 g	55	0	32
Cantaloupe melon, raw, ¼ small, 200 g	65	0	10
Cherries, 20, 80 g	22	0	10
Fruit cocktail, canned in natural juice, 125 g	55	0	15
Grapefruit juice, unsweetened, 250 ml	48	0	16
Grapefruit, raw, ½ medium, 100 g	25	0	5
Grapes, green, 100 g	46	0	15

Food	G.I.	Fat	CHO
		(grams per serving)	
Fruits and fruit products (*cont.*)			
Kiwifruit, 1 raw, peeled, 80 g	52	0	8
Lychee, canned and drained, 7, 90 g	79	0	16
Mango, 1 small, 150 g	55	0	19
Orange, 1 medium, 130 g	44	0	10
Orange juice, 250 ml	46	0	21
Pawpaw, ½ small, 200 g	58	0	14
Peach, fresh, 1 large, 110 g	42	0	7
canned, natural juice, 125 g	30	0	12
canned, heavy syrup, 125 g	58	0	19
canned, light syrup, 125 g	52	0	18
Pear, fresh, 1 medium, 150 g	38	0	21
canned in pear juice, 125 g	44	0	13
Pineapple, fresh, 2 slices, 125 g	66	0	10
Pineapple juice, unsweetened, canned, 250 ml	46	0	27
Plums, 3–4 small, 100 g	39	0	7
Raisins, 40 g	64	0	28
Sultanas, 40 g	56	0	30
Watermelon, 150 g	72	0	8
Gluten-free bread, 1 slice, 30 g	90	1	14

Food	G.I.	Fat	CHO
		(grams per serving)	
Glutinous rice, white, steamed, 1 cup, 174 g	98	0	37
Gnocchi, cooked, 145 g	68	3	71
Grapefruit juice, unsweetened, 250 ml	48	0	16
Grapefruit, raw, ½ medium, 100 g	25	0	5
Grape Nuts™ cereal, ½ cup, 58g	71	1	47
Grapes, green, 100 g	46	0	15
Green gram dhal, 100 g	62	4	10
Green gram, soaked and boiled, 120 g	38	1	18
Green pea soup, canned, ready to serve, 220 ml	66	1	22
Hamburger bun, 1 prepacked, 50 g	61	3	24
Haricot (navy beans), boiled, 90 g	38	0	11
Honey & Oat Bread (Vogel's™), 1 slice, 40 g	55	3	17
Honey, 1 tablespoon, 20 g	58	0	16
Ice cream, full fat, 2 scoops, 50 g	61	6	10
Ice cream, low-fat, 2 scoops, 50 g	50	2	13
Jelly beans, 5, 10 g	80	0	9
Kidney beans, boiled, 90 g	27	0	18

Food	G.I.	Fat	CHO
		(grams per serving)	
Kidney beans, canned and drained, 95 g	52	0	13
Kiwifruit, 1 raw, peeled, 80 g	52	0	8
Lactose, pure, 10 g	46	0	10
Lentil soup, canned, 220 ml	44	0	14
Lentils, green and brown, dried, boiled, 95 g	30	0	16
Lentils, red, boiled, 120 g	26	1	21
Light rye bread, 1 slice, 50 g	68	1	23
Linguine pasta, thick, cooked, 180 g	46	1	56
Linguine pasta, thin, cooked, 180 g	55	1	56
Linseed rye bread, 1 slice, 50 g	55	5	21
Lucozade ™, original, 1 bottle, 300 ml	95	<1	56
Lungkow bean thread, 180 g	26	0	61
Lychee, canned and drained, 7, 90 g	79	0	16
Macaroni cheese, packaged, cooked, 220 g	64	24	30
Macaroni, cooked, 180 g	45	1	56
Maize			
Cornmeal, wholegrain, 40 g	68	1	30

Food	G.I.	Fat	CHO
		(grams per serving)	
Maize (*continued*)			
Sweet corn, canned and drained, 80 g	55	1	16
Maltose (maltodextrins), pure, 10 g	105	0	10
Mango, 1 small, 150 g	55	0	19
Mars Bar™, 60 g	68	11	41
Melba toast, 4 squares, 30 g	70	1	19
Milk, full fat, 250 ml	27	10	12
Milk, skimmed, 250 ml	32	0	13
chocolate flavoured, 250 ml	34	3	23
Milk, sweetened condensed, ½ cup, 160 g	61	15	90
Milk Arrowroot biscuits, 2,16 g	63	2	13
Millet, cooked, 120 g	71	0	12
Mini Wheats™ (whole wheat) breakfast cereal, 30 g	58	0	21
Morning Coffee biscuits, 3, 18 g	79	2	14
Muesli bars with fruit, 30 g	61	4	17
Muesli, breakfast cereal			
toasted, 60 g	43	9	33
non-toasted, 60 g	56	6	32
Muffins			
Apple, 1 muffin, 80 g	44	10	44

Food	G.I.	Fat	CHO
		(grams per serving)	
Muffins (*continued*)			
Bran, 1 muffin, 80 g	60	8	34
Blueberry, 1 muffin, 80 g	59	8	41
Mung bean noodles, 1 cup, 140 g	39	0	35
Noodles, 2-minute, 85 g packet, cooked	46	16	55
Noodles, rice, fresh, boiled, 1 cup 176 g	40	0	44
Oat bran, raw, 1 tablespoon, 10 g	55	1	7
Oatmeal biscuits, 3 biscuits, 30 g	54	6	19
Orange, 1 medium, 130 g	44	0	10
Orange juice, 250 ml	46	0	21
Orange squash, diluted, 250 ml	66	0	20
Parsnips, boiled, 75 g	97	0	8
Pasta			
Capellini, cooked, 180 g	45	0	53
Fettucini, cooked, 180 g	32	1	57
Gnocchi, cooked, 145 g	68	3	71
Noodles, 2-minute, 85 g packet, cooked	46	16	55
Linguine, thick, cooked, 180 g	46	1	56
Linguine, thin, cooked, 180 g	55	1	56

Food	G.I.	Fat	CHO
		(grams per serving)	
Pasta (*continued*)			
Macaroni cheese, packaged, cooked, 220 g	64	24	30
Macaroni, cooked, 180 g	45	1	56
Noodles, mung bean, 1 cup, 140 g	39	0	35
Noodles, rice, fresh, boiled, 1 cup, 176 g	40	0	44
Ravioli, meat-filled, cooked, 220 g	39	11	30
Rice pasta, brown, cooked, 180 g	92	2	57
Spaghetti, white, cooked, 180 g	41	1	56
Spaghetti, wholemeal, cooked, 180 g	37	1	48
Spirale, durum. cooked, 180 g	43	1	56
Star pastina, cooked, 180 g	38	1	56
Tortellini, cheese, cooked, 180 g	50	8	21
Vermicelli, cooked, 180 g	35	0	45
Pastry, flaky, 65 g	59	26	25
Pawpaw, raw, ½ small, 200 g	58	0	14
Pea and ham soup, canned, 220 ml	66	2	13

Food	G.I.	Fat	CHO
		(grams per serving)	
Peach, fresh, 1 large, 110 g	42	0	7
canned, natural juice, 125 g	30	0	12
canned, heavy syrup, 125 g	58	0	19
canned, light syrup, 125 g	52	0	18
Peanuts, roasted, salted, 75 g	14	40	11
Pear, fresh, 1 medium, 150 g	38	0	21
canned in pear juice, 125 g	44	0	13
Peas, green, fresh. frozen, boiled, 80 g	48	0	5
Peas, dried, boiled, 70 g	22	0	4
Pineapple, fresh, 2 slices, 125 g	66	0	10
Pineapple juice, unsweetened, canned, 250 g	46	0	27
Pinto beans, canned, 95 g	45	0	13
Pinto beans, soaked, boiled, 90 g	39	0	20
Pitta bread, 1 piece, 65 g	57	1	38
Pizza, cheese and tomato, 2 slices, 230 g	60	27	57
Plums, 3–4 small, 100 g	39	0	7
Popcorn, low-fat (popped), 20 g	55	2	10
Porridge (made with water), 245 g	42	2	24
Potatoes			
French Fries, fine cut, small serving, 120 g	75	26	49

Food	G.I.	Fat	CHO
		(grams per serving)	
Potatoes (*continued*)			
instant potato	83	1	18
new, peeled, boiled, 5 small (cocktail), 175 g	62	0	23
new, canned, drained, 5 small, 175 g	61	0	20
pale skin, peeled, boiled, 1 medium, 120 g	56	0	16
pale skin, baked in oven (no fat), 1 medium, 120 g	85	0	14
pale skin, mashed, 120 g	70	0	16
pale skin, steamed, 1 medium, 120 g	65	0	17
pale skin, microwaved, 1 medium, 120 g	82	0	17
potato crisps, plain, 50 g	54	16	24
Potato crisps, plain, 50 g	54	16	24
Pound cake, 1 slice, 80 g	54	15	42
Pretzels, 50 g	83	1	22
Puffed crispbread, 4 wholemeal, 20 g	81	1	15
Puffed wheat breakfast cereal, 30 g	80	1	22
Pumpernickel bread, 2 slices	41	2	35

Food	G.I.	Fat	CHO
		(grams per serving)	
Pumpkin, peeled, boiled, 85 g	75	0	6
Raisins, 40 g	64	0	28
Ravioli, meat-filled, cooked, 20 g	39	11	30
Rice			
Basmati, white, boiled, 180 g	58	0	50
Glutinous, white, steamed, 1 cup, 174 g	98	0	37
Instant, cooked, 180 g	87	0	38
Rice bran, extruded, 1 tablespoon, 10 g	19	2	3
Rice cakes, 2, 25 g	82	1	21
Rice Krispies™, breakfast cereal, 30 g	82	0	27
Rice noodles, fresh, boiled, 1 cup, 176 g	40	0	44
Rice pasta, brown, cooked, 180 g	92	2	57
Rice vermicelli, cooked, 180 g	58	0	58
Rich Tea biscuits, 2, 20	55	3	16
Rye bread, 1 slice, 50 g	65	1	23
Ryvita™ crackers, 2 biscuits, 20 g	69	1	16
Sausages, fried, 2, 120 g	28	21	6
Semolina, cooked, 230 g	55	0	17
Shortbread, 2 biscuits, 30 g	64	8	19

Food	G.I.	Fat	CHO
		(grams per serving)	
Shredded wheat breakfast cereal, 25 g	67	0	18
Soda crackers, 3 biscuits, 25 g	74	4	17
Soft drink, Coca Cola™, 1 can, 375 ml	63	0	40
Soft drink, Fanta™, 1 can, 375 ml	68	0	51
Soups			
Black bean soup, 220 ml	64	2	82
Green pea soup, canned, ready to serve, 220 ml	66	1	22
Lentil soup, canned, 220 ml	44	0	14
Pea and ham soup, 220 ml	60	2	13
Tomato soup, canned, 220 ml	38	1	15
Sourdough rye bread, 1 slice, 50 g	57	2	23
Soya beans, canned, 100 g	14	6	12
Soya beans, boiled, 90 g	18	7	10
Spaghetti, white, cooked, 180 g	41	1	56
Spaghetti, wholemeal, cooked, 180 g	37	1	48
Special K™, 30 g	54	0	21
Spirale pasta, durum, cooked, 180 g	43	1	56
Split pea soup, 220 ml	60	0	6
Split peas, yellow, boiled, 90 g	32	0	16

Food	G.I.	Fat	CHO
		(grams per serving)	
Sponge cake plain, 1 slice, 60 g	46	16	32
Sports drinks			
Gatorade, 250 ml	78	0	15
Isostar, 250ml	70	0	18
Stoned wheat thins, crackers, 5 biscuits, 25 g	67	2	17
Sucrose, 1 teaspoon	65	0	5
Sultana Bran™, 45 g	52	1	35
Sultanas, 40 g	56	0	30
Sustain™, 30 g	68	1	25
Swede, peeled, boiled, 60 g	72	0	3
Sweet corn, 85 g	55	1	16
Sweet potato, peeled, boiled, 80 g	54	0	16
Sweetened condensed milk, ½ cup, 160 g	61	15	90
Taco shells, 2, 26 g	68	6	16
Tapioca pudding, boiled with milk, 250 g	81	10.5	51
Tapioca, steamed 1 hour, 100 g	70	6	54
Tofu frozen dessert (non-dairy), 100 g	115	1	13
Tomato soup, canned, 220 ml	38	1	15
Tortellini, cheese, cooked, 180 g	50	8	21

Food	G.I.	Fat	CHO
		(grams per serving)	
Vanilla wafer biscuits, 6, 30 g	77	5	21
Vermicelli, cooked, 180 g	35	0	45
Waffles, 25 g	76	3	9
Water biscuits, 5, 25 g	78	2	18
Watermelon, 150 g	72	0	8
Weetabix™ breakfast cereal, 2 biscuits, 30 g	69	1	19
Wheatmeal biscuits, 2, 16 g	62	2	12
White bread, wheat flour, 1 slice, 30 g	70	1	15
Wholemeal bread, wheat flour, 1 slice, 35 g	69	1	14
Yakult, 65 ml serve	46	0	11
Yam, boiled, 80 g	51	0	26
Yoghurt			
low-fat, fruit, 200 g	33	0	26
low-fat, artificial sweetener, 200 g	14	0	12

WHERE TO GO FOR HELP AND FURTHER INFORMATION

British Diabetic Association
10 Queen Anne Street
London W1M 0BD
Telephone: 0207 323 1531

British Dietetic Association
5th Floor, Elizabeth House
22 Suffolk Street Queensway
Birmingham B1 1LS
Telephone: 0121 616 4900

Irish Nutrition & Dietetic Institute
Dundrum Business Centre
Frankfort Dundrum
Dublin 14
Ireland
Telephone: (1)298 7466

ABOUT THE AUTHORS

Kaye Foster-Powell, an accredited practising dietitian-nutritionist, has extensive experience in diabetes management and has researched practical applications of the glycaemic index. She is the senior dietitian at Wentworth Area Diabetes Service and conducts a private practice in the Blue Mountains, New South Wales. Her most recent book is the best-selling *Glucose Revolution*, published by Hodder & Stoughton.

Dr Jennie Brand-Miller, is Associate Professor of Human Nutrition in the Human Nutrition Unit at the University of Sydney. She is a world authority on the glycaemic index of foods and its applications to health. Her most recent book is the best-selling *Glucose Revolution*, published by Hodder & Stoughton.

Dr Stephen Colagiuri, Director of the Diabetes Centre and Head of the Department of Endocrinology, Metabolism and Diabetes at the Prince of Wales Hospital in Randwick, New South Wales, has published extensively on the importance of carbohydrate in the diet of people with diabetes. His most recent book was the best-selling *Glucose Revolution*, published by Hodder & Stoughton.